School Publishing

# Sound All Around

PATHFINDER EDITION

By Rebecca L. Johnson

CONTENTS

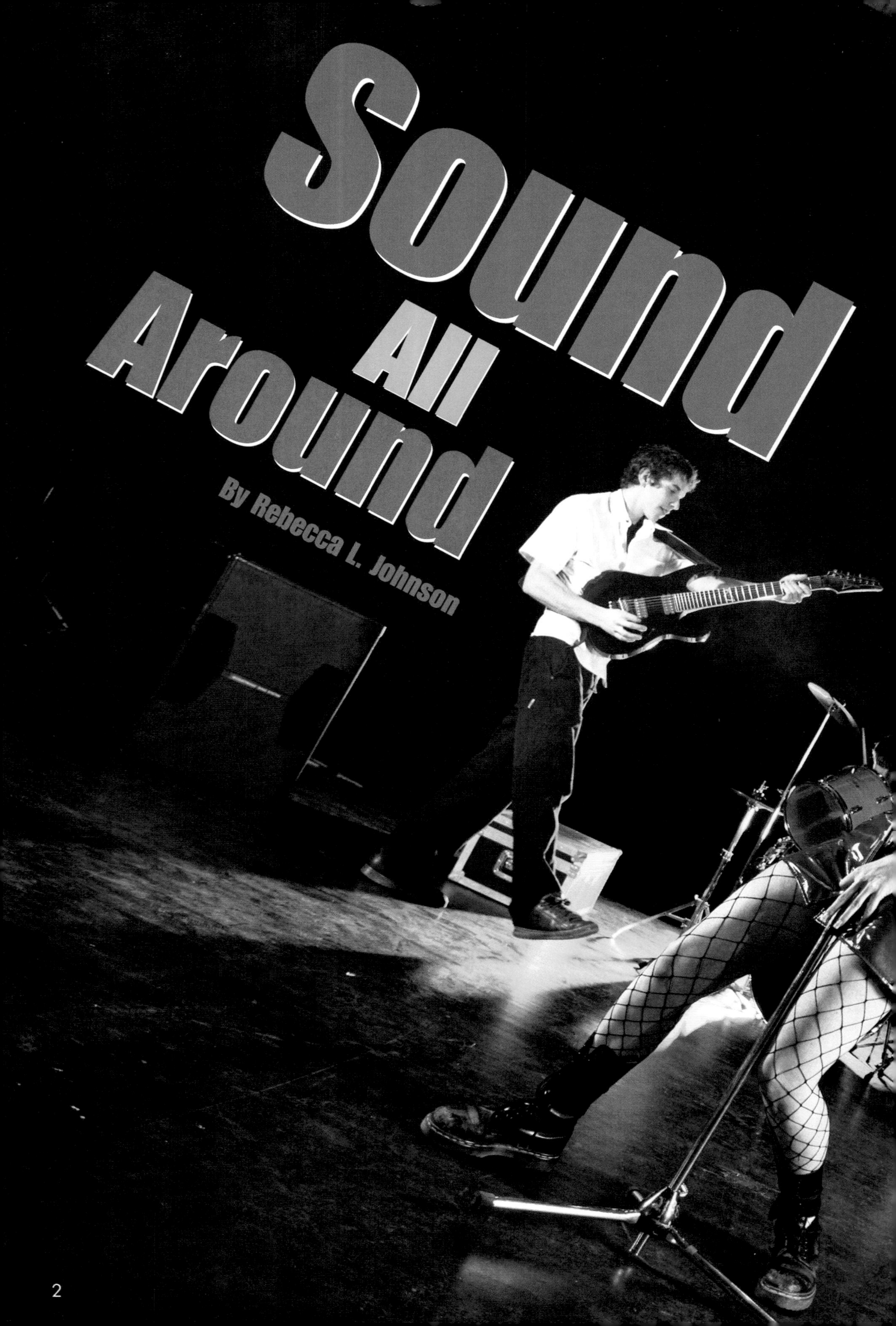
Sound
All
Around
By Rebecca L. Johnson

The band members stride onto the stage. Spotlights light the stage. The clapping quiets down and the music begins. Suddenly, you're surrounded by sound that makes you want to dance!

Sound is all around you, all the time. Just listen. Do you hear people talking? Is a computer humming? Can you hear traffic outside?

Sounds can seem very different. But all sounds are a form of **mechanical energy**. Sounds get their start when objects move quickly back and forth. These back-and-forth movements are **vibrations**.

When a musician plucks a guitar string, it vibrates. The string's vibrations cause air particles near the string to vibrate. Those particles cause other particles to vibrate, too. In this way, the vibration travels through the air.

## The Wave

The vibration moves like an invisible wave. In fact, it's called a **sound wave**. Sound waves spread out in all directions from their source. And they don't just travel through air. Sound waves move through water and other liquids. They move through solids, too.

At a rock concert, all the musical instruments make sounds by vibrating. Guitar strings vibrate when guitar players strum them. Drums vibrate when the drummer hits them. These vibrations create sound waves. Singers use their vocal cords in their throat to create sound waves. When vocal cords vibrate, they make sound waves. Your vocal cords vibrate when you talk, shout, and cheer. That's how you make your own sound waves.

**Cool Vibrations.** Long ago, guitar strings used to be made from animal intestines. Today they're made from nylon and metals such as bronze, nickel, and steel.

## Hearing Sounds

How do you hear sounds? Sound waves move through the air and reach your ears. The outside part of each ear is shaped kind of like a funnel. It helps sound waves flow through the ear canal to the eardrum, which is deeper inside. When sound waves hit the eardrum, it vibrates.

Vibrations from the eardrum move on to a set of tiny ear bones. The bones pass the vibrations on to a liquid-filled tube. There, nerves pick up the vibrations. They send information about them to your brain. Your brain interprets this information as sound.

It's also possible to feel sounds. At a loud concert, you can feel the vibrations from sound waves in your whole body.

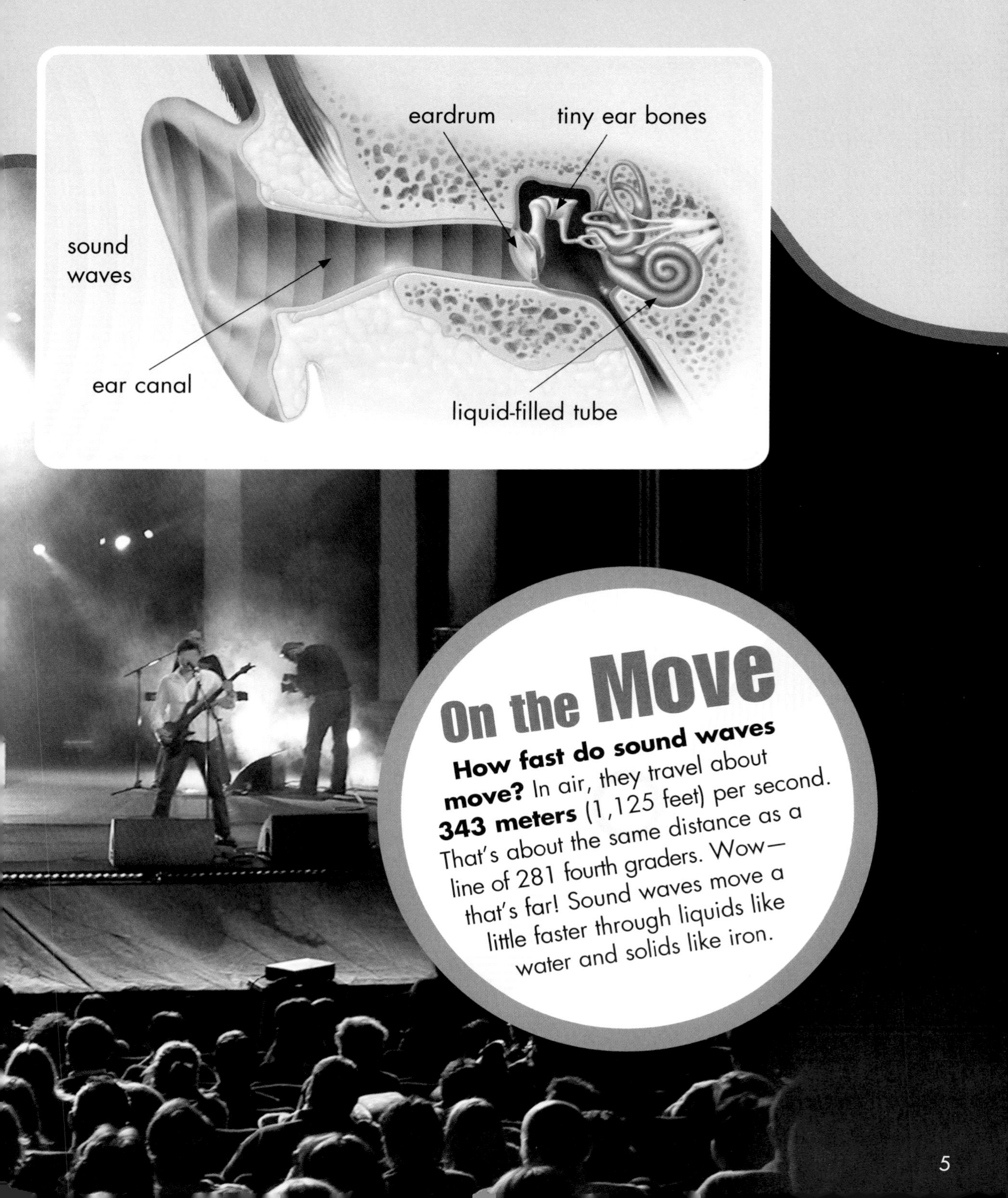


### On the Move

**How fast do sound waves move?** In air, they travel about **343 meters** (1,125 feet) per second. That's about the same distance as a line of 281 fourth graders. Wow—that's far! Sound waves move a little faster through liquids like water and solids like iron.

**Making Waves.** When you play the lowest notes on a keyboard, the slower vibrations make a wavy sound.

## Highs and Lows

Sounds have different qualities. Some sounds are high and others are low. A sound's highness or lowness is called **pitch**. High-pitch sounds are produced by fast vibrations. The top keys on a real or digital piano keyboard make high-pitch sounds. Low-pitch sounds are produced by slow vibrations. The bottom keys on the keyboard make low-pitch sounds.

Scientists call a sound's pitch its **frequency**. Frequency tells how many vibrations per second a sound wave has. A sound with a high frequency has more vibrations per second than a sound with a low frequency. High-frequency sounds have a high pitch. Low-frequency sounds have a low pitch.

## Sound? What Sound?

People can hear many, but not all, frequencies of sound. Dogs, cats, and some other animals hear sounds with frequencies higher than people can hear. When someone blows a dog whistle, a dog can hear the sound it makes. Yet most people can't. The whistle makes sound waves that have a higher frequency than human ears can hear.

Whales make some extremely low-pitch sounds. These sounds have frequencies that are lower than people can hear. The sounds can travel very far.

## Turn It Up!

Sounds can be loud, soft, or anything in between. A sound's loudness depends on how much energy the sound wave has.

A loud sound is produced by big vibrations. A loud sound has a lot of energy. Hitting a drum very hard produces a loud sound.

A soft sound is produced by small vibrations. A soft sound has only a little energy. Raindrops gently hitting a windowpane produce a soft sound.

Now you know a lot more about sounds. So listen again. What high-pitch, low-pitch, loud, and soft sounds do you hear? And what is vibrating to send those sounds toward your ear?

**Long Distance Call.** Whales use the sounds they make to keep in touch in the ocean.

## Wordwise

**frequency:** how many vibrations per second a sound wave has

**mechanical energy:** energy something has due to its motion or position

**pitch:** the highness or lowness of a sound

**sound wave:** a wave of energy that we hear as sound

**vibration:** very fast back-and-forth movements

# How LOUD Is That?

Everyone knows that rock concerts are loud. That's especially true if you're sitting close to the stage. But just how loud are they?

Scientists use units called decibels to measure the energy of a sound. The greater the number of decibels, the more energy a sound has. And the more energy it has, the louder it is.

A sound that measures 0 decibels is so soft it can barely be heard. A sound that measures 10 decibels is ten times louder than the 0-decibel sound. A sound that measures 20 decibels is a hundred times louder than a 0-decibel sound. A sound that measures 30 decibels is a thousand times louder, and so on.

People can hear sounds that range from about 0.1 decibels to 150 decibels. A 160-decibel sound breaks human eardrums. But even sounds over 85 decibels are loud enough to hurt hearing. How can you protect your ears? Avoid being around loud sounds, especially for hours at a time.

The table shows some common sounds. Check out how many decibels each sound has.

| Sound | Almost total silence | Breathing | Soft whisper | Refrigerator humming | Normal conversation | Vacuum cleaner |
| --- | --- | --- | --- | --- | --- | --- |
| Decibels | 0 | 10 | 30 | 40 | 60–70 | 80 |

| Lawn mower | Personal stereo, maximum level | Motorcycle | Chain saw | Front row of rock concert | Jet engine up close |
| --- | --- | --- | --- | --- | --- |
| 85–110 | 90 | 105 | 110 | 115 | 150 |

# Going, Going, Gone?

Can you imagine inventing a sound? That's just what a security company in Britain did. They created a very high-pitch sound.

The company called their special sound the Mosquito. They sold the sound to store owners. The owners played it to bother young people who stood around outside their stores. The sound was so annoying that the young people would walk away. Most adults didn't notice. They couldn't hear the Mosquito.

Why can kids hear sounds that most adults can't? Hearing changes as people get older. Nearly everyone slowly loses the ability to hear really high-pitch sounds. The loss is caused by changes that take place deep inside the ear. It starts when a person is about 18 years old. It is usually greater in men than women.

The bad news about this hearing loss is that it happens to nearly everyone. The good news is that it takes a long time. Most adults don't even notice the change in their hearing until they are 60 years old or older. There is more good news, too. Scientists recently found some genes that may be responsible for this hearing loss. Someday they may be able to control those genes in people and prevent the problem. So stay tuned!

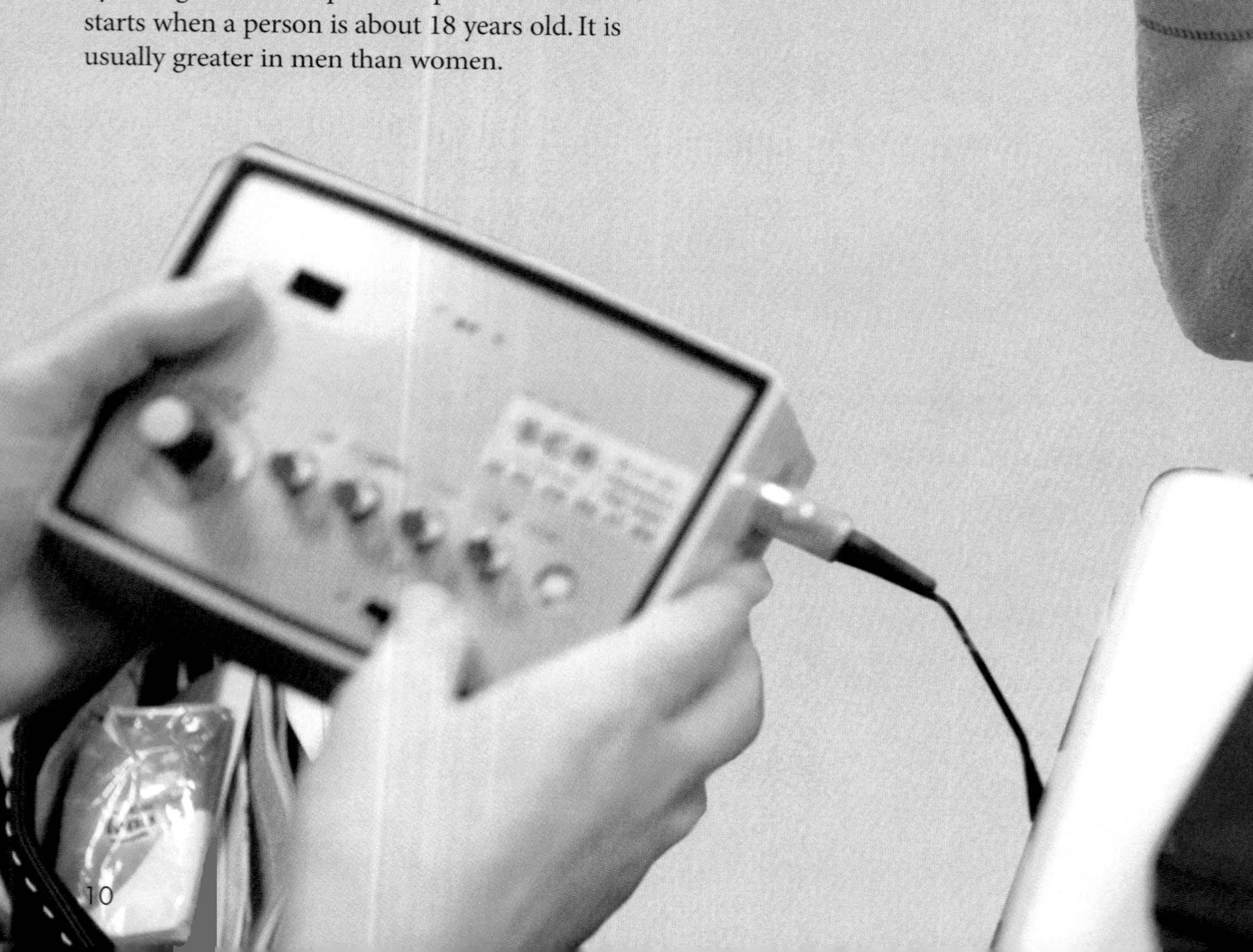

# Sound All Around

**Sound off about what you've learned by answering the questions below.**

1. How are sounds and vibrations connected?
2. Where can sound waves travel?
3. Explain what sound waves do when they reach your ears.
4. What is pitch?
5. Why can kids hear some sounds that most adults can't?